The Orchids of Lovell Hollow

and

Heartsong Retreat Center

AN ENUMERATION AND DESCRIPTION OF
THE WILD ORCHIDS OF LOVELL HOLLOW IN
NEWTON COUNTY, ARKANSAS.

Written and photographed
by Olin Karch

Text and photographs © Copyright 2007 Olin Karch
All rights reserved
ISBN 978-1-105-78347-0

Published by the Author
HC 62 Box 826, Deer, Arkansas, 72628

Printed by Lulu.com

Cover photo: slender ladies'-tresses at Heartsong Shrine

Introduction

At the mention of 'wild' orchids many people react with confusion. They may have a concept of orchids as exotic, delicate, showy, and expensive hothouse plants like the *Cattleya*, shown here, or are suitable only for climates like Hawaii.

They're right.

Sometimes.

Orchids are also hardy alpine wildflowers in cold climates.

Sometimes.

Orchids are the largest family of plants, with over 30,000 species, and found native on every continent except Antarctica.

Orchids are found wild in every state in the Union. In fact, Hawaii has fewer species of native orchids than any other state. It has only three species, and those are small and inconspicuous. All those spectacular orchids they grow are imported from other countries.

In numbers of species and their beauty, Arkansas blows Hawaii out of the water. There have been over 40 kinds of orchids documented in Arkansas.

Another popular misconception about orchids is that they live as parasites in steamy jungles. Although many orchids grow in trees, those are not parasites. Instead, they are 'epiphytes' meaning they live in trees, but they use their thick corky roots to anchor themselves and to soak up water during the rainy season, but do not actually obtain any nourishment from the host tree. Many orchids store water in bulbous stems and leaves, as they may have to endure months without any water during the dry season.

Orchids often cannot survive the hot steamy conditions people imagine them to need. Most do best with temperatures in the 70's and many make good houseplants. In fact some orchids such as *Cymbidiums* may not bloom if they are not given cool growing conditions.

And even though 'orchid' is a color, orchids come in every color of the rainbow.

OK, OK, but what makes them all orchids?

Well, orchids are **monocots**, with an **inferior ovary** and three **sepals** and three **petals**, one of which is usually modified into a showy **lip**, or **labellum**. The male and female reproductive organs are modified into a single structure called a **column**, where the pollen is born in discrete masses called **pollinia**. The **stigma** is modified with a sticky pad strong enough to wrench the pollinia from the backs of unsuspecting insects, insuring cross-pollination of an entire pod of often over a million seeds at one time.

There, aren't you sorry you asked?

Acknowledgments

While this booklet has largely been written for my own entertainment, there are several people who deserve a special thank you for their part in guiding me, or the contents, to final form. My high school biology teacher, the late Dr. Earl N. Newman triggered my interest in orchids, and my mother nurtured it, and my orchid collection while I was absent at college. Dominic Fabis first led me to this location where I now live, and my neighbor, Don Kitz helped me make contacts to acquire the property. Don also led my first hike into the wilderness, on which I spotted two species of orchids, even though it was early winter. Dr. George Johnson verified my identifications and provided Arkansas range maps for my finds, and Dr. Doug Martin proofread and offered suggestions in wording. Paul Martin Brown generously checked my US range maps and made corrections and updates. My partner, Steven Wilgus kept me fed and clothed and sane, here in the wilderness while writing it. All have my profuse gratitude for their help.

Preface

I became fascinated with orchids when I was in high school, and my biology teacher assigned me the task of finding how to keep alive an orchid, *Epidendrum nocturnum* that had been purchased as a classroom example of an epiphyte. It died, but being a lover of the exotic and unusual, I was determined to grow orchids of my own. Not only was that the start of a lifelong passion for growing these exotic plants, but finding out there might be native orchids where I lived, whetted my interest in botany, which became my major in college.

While I have had the good fortune to visit the tropics on several occasions and seen exotic orchid species in their native habitat in Central and South America, I still have a particular fascination for the often-unassuming native terrestrial species in the U.S. I had occasionally encountered wild orchids when I lived in Oklahoma, California, and Kansas, but I never found more than one or two species in an area.

The first time I visited northwest Arkansas as an adult I encountered two species of wild orchids on a hike to the Buffalo River and I was hooked. Each time I went back I found another species. When the opportunity arose for me to be able to purchase 25 acres of woodland here, I jumped at the chance. It had been a fantasy from high school to own land in the Ozarks, and this was a golden opportunity to live in an area, which was not only beautiful, but also botanically rich.

I set a goal of seeing how many species I could find within an hour's walk of my cabin. As of this writing I have found 15 species or varieties of wild orchids along the path between my cabin and the river.

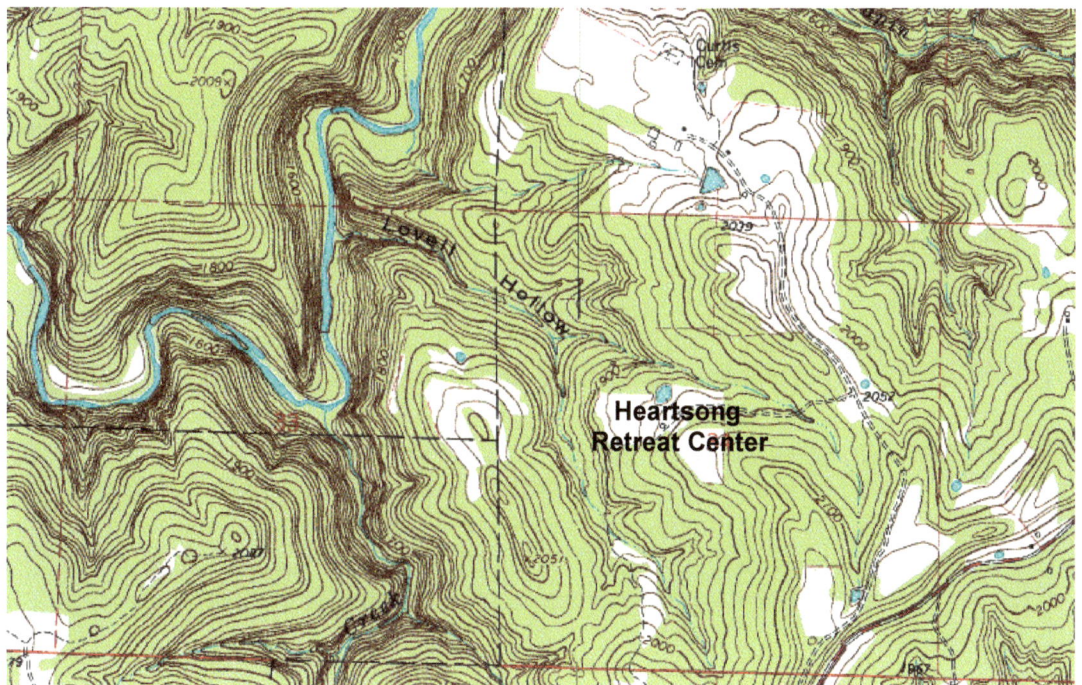

This area, known as Lovell Hollow, extends from the Buffalo River, in the Upper Buffalo Wilderness Area, at about 1500 feet elevation, toward the top of the plateau at about 2000 feet.

At the level of the river are rich hardwood forests on level ground. The trail to the top follows an old logging road and passes through dryer woods and crosses a few seepages or temporary streams. At the top of the ridge are artificially cleared areas as well as more dry forests.

The forests at the river bottoms are the richest in orchid species, but there are species adapted to each habitat, even regularly mowed lawns. Orchids may be found all along the trail to the river, but the most abundant and conspicuous places which will be referred to are the bottomland forest, at about 1500 feet, the middle forest at about 1800 feet, and the ridge at about 2000 feet.

Some species, like the ladies'-tresses prefer open habitats such as mowed lawns or cleared paths, and some like the lady's-slippers are to be found only in the bottomlands. A few species, such as the downy rattlesnake orchid, the coralroots, and the three birds orchid may be encountered in any of the wooded areas. At one spot in the bottomlands, I encountered 7 species, in an area no greater than 10 feet square.

The orchids described here all have specific blooming seasons, which vary with location. For example, *Spiranthes vernalis*, which is considered spring blooming, but may be found in bloom somewhere in North America, almost any time in the year. In Arkansas it blooms from May to July, but at this elevation, it blooms in July. For the purposes of this booklet, I have listed the blooming season according to when I have found the plants blooming in Lovell Hollow.

Format of this book

For each species I have tried to include common and scientific names, photos of the flowers and the plant habit, as well as give information about its range in the United States, where it may be found in Arkansas, and where it may be found in or near Lovell Hollow. The U.S. maps show ranges of the species in green, or in red where the species is on a state concern list (threatened, endangered, etc.) The maps of Arkansas show verified county records in green and additional unverified reports in blue. Blooming seasons vary considerably throughout the range of a species, so the blooming season given in this booklet refers to when it might be expected at this location. Where medicinal uses or folk traditions are known, they are included at the end of the write-up. Common names as given here are lower case except where a proper name is involved, and scientific names follow standard protocol with the genus capitalized and species lower case, and both names italicized.

Adam-and-Eve or putty-root

Aplectrum hyemale

Blooms in May

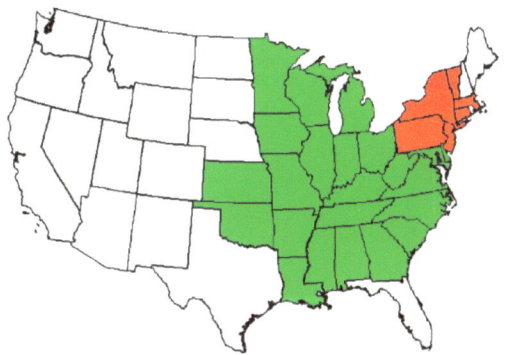

Found from Minnesota and Ontario east to Vermont and Massachusetts, south to Arkansas and Georgia, it is listed as a species of conservation concern in several New England states.

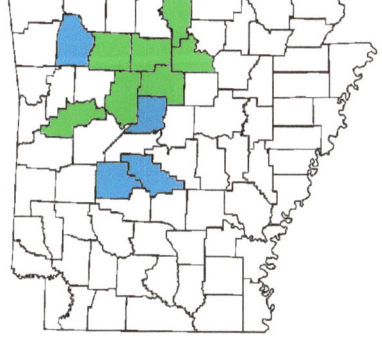

In Arkansas it is found in the Ozarks and Ouachita Mountains.

This orchid grows in the leaf litter of hardwood forests. Here it can be found anywhere from the top of the ridge to the river bottoms.

The root of this orchid grows a new section each year and retains the previous growth, forming a pair of bulb-like swellings connected by a narrow root, giving a sort of dumbbell appearance. The paired appearance of the root is said to be the source of the name Adam-and-Eve. The root is said to form a glue when mashed, accounting for the name of 'putty-root.'

The species name, *hyemale* means 'in winter' and refers to the evergreen leaves of the plant, which can be seen on the forest floor in the winter. The plants take advantage of the deciduous forest by emerging in October and remaining green through the winter. The leaves wither shortly before the blooms appear in spring.

The flower stalk stands up to two feet tall, and bears about 15 flowers, each white and brown to purple, about half an inch long.

Indians used the roots in a poultice for boils, and the roots have been used in a tea for bronchitis. Cherokee deer hunters carried the root and when killing a deer, put a bit of chewed root into the wound to assure that the animal would be found to be unusually fat when skinned. If a baby was not thriving, the root was used in a decoction to bath the infant, which was believed to fatten it.

autumn coralroot

Corallorhiza odontorhiza

var. *odontorhiza*

Blooms in August and September

This orchid is found all over the eastern half of the United States down through Mexico, Belize, Guatemala, and Honduras. In Arkansas it is found in scattered locations in mostly mountainous areas. It is a species of conservation concern in many states.

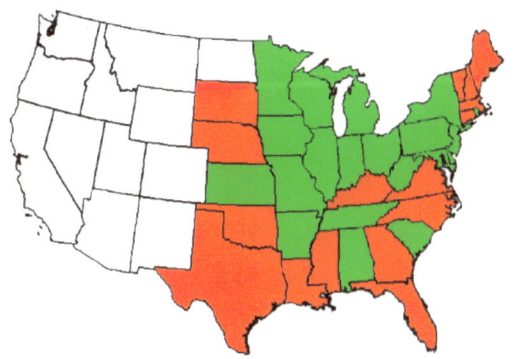 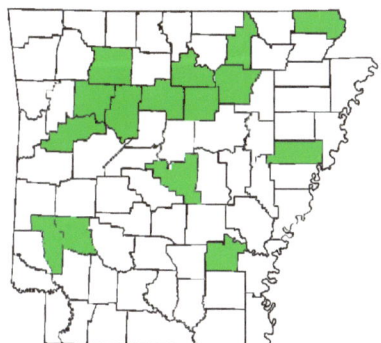

Here it is found in woods from the top of the ridge to the river bottoms.

The coralroot orchids are an interesting group because they are mycotrophic. This means that they have no chlorophyll and do not manufacture their own food. They 'feed' instead digesting leaf litter with the help of fungi in their roots. Because of their nature, coralroots can survive underground and may not be seen for years at a time, needing only to emerge for flowering and seed production.

The flowers of the autumn coralroot are inconspicuous, and easily overlooked. This variety is cleistogamous, meaning the flowers self-pollinate, may not actually open, and the lip may be undeveloped.

Occasionally plants have been found locally which are chasmogamous, that is to say, the flowers open fully and unlike the typical variety of the species, do not self-pollinate while still budded. The lip is fully developed and the sepals open to show the petals and lip.

While there is a specific variety named "**Pringle's autumn coralroot**" found in the Great Lakes region which regularly bears chasmogamous flowers, the plants found here are probably only exhibiting individual or environmental variation. One plant observed near Alum Cove appeared to bear both types of flowers growing from the same base.

spring coralroot or Wister's coralroot

Corallorhiza wisteriana

Blooms in May

This orchid ranges over most of the U. S,
Except for most western and northern states.
It is listed as a species of conservation concern
in many of the states.

In Arkansas it is found scattered throughout the
State.

Here it can be found in forested areas,
most abundant in the bottomlands near the river.

Like the autumn coralroot, this plant has no
leaves and no chlorophyll, and lives on digested
leaf litter with the help of a fungus. It can survive
for years underground and only sends
up stems to flower and produce seeds.

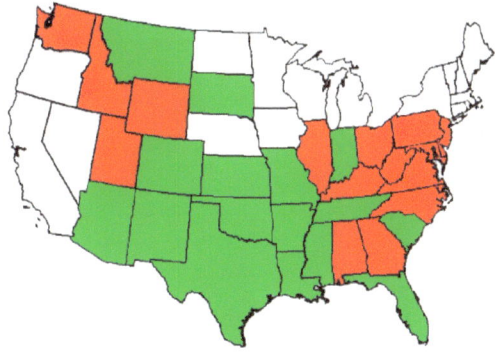
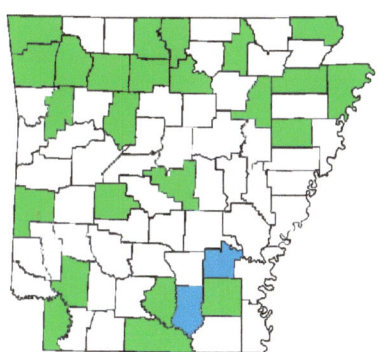

spring coralroot, *Corallorhiza wisteriana*

large yellow lady's-slipper

Cypripedium parviflorum

var. pubescens

Blooms in April and May

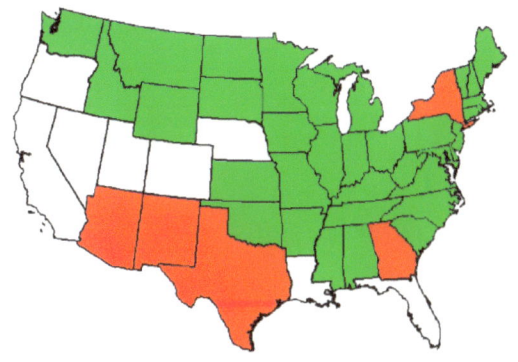

Found from Alaska to Newfoundland, and south to Arizona and Georgia. It is on lists of conservation concern in several states.

This is one of four kinds of lady's-slipper orchids found in Arkansas, and the commonest of the four.

These orchids are found here in fair abundance scattered in the woods along the bottomlands.

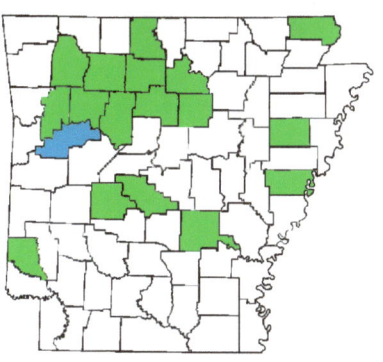

These plants stand up to two feet tall and bear one or two flowers, which may measure up to 6 inches from top to bottom.

The size and color of this variety may vary, as seen in the all yellow flower on the left above. Another variety is found in this county, which has smaller flowers and darker brown on the sepals and petals, and the two varieties may intergrade with one another. Our orchids appear to all be of the variety *pubescens* but the variety *parviflorum* is probably also found nearby.

These orchids were once used medicinally as a sedative for various mental disorders and apparently what is now known as PMS. Such use, as well as digging for gardens has wiped out this species in many areas and it is under threat from loss of habitat as well.

showy orchis

Galearis spectabilis

Blooms in May

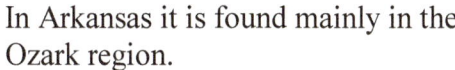

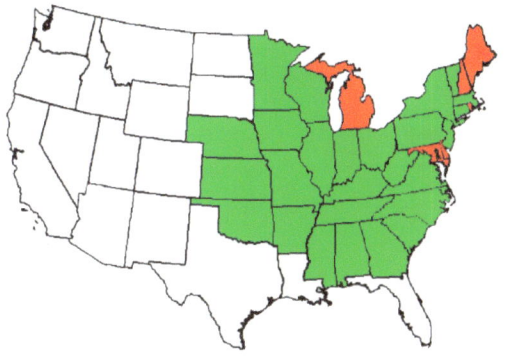

The showy orchis is found from Minnesota to New Brunswick, and south to Oklahoma and Georgia. It is on the lists of species of conservation concern in several northern states.

In Arkansas it is found mainly in the Ozark region.

Here it may be found in the rich woods of the river bottoms, particularly on the far side of the river.

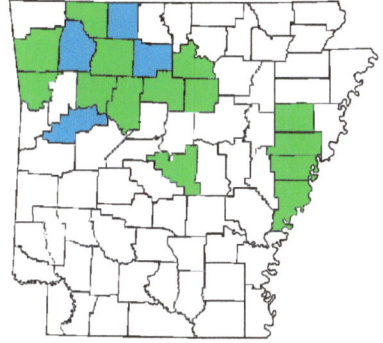

This orchid typically has pink or purple sepals and petals, which cup together to form a hood over the white lip, but may vary from all purple to all white.

9

downy rattlesnake orchid

Goodyera pubescens

Blooms in August and September

This orchid ranges through the eastern half of the United States, and may be quite common locally.

In Arkansas it is found in the Ozarks and Ouachita Mountains.

Here it may be found in wooded areas all the way from the top of the ridge to the bottomlands of the river.

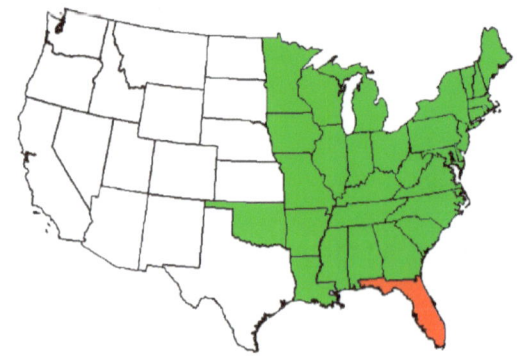

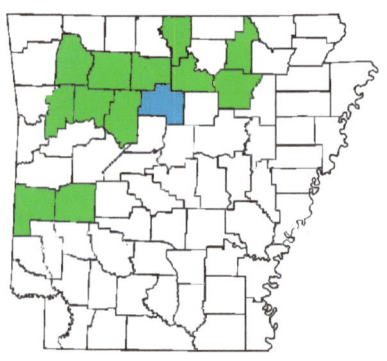

The small white flowers of the downy rattlesnake orchid, arranged in a tight spiral, are attractive, if inconspicuous. The foliage, which is a very attractive dark velvety green with light green netting, is evergreen, and can be found easiest in the winter when it takes advantage of the deciduous hardwood forest to get its share of the sunlight.

The netted pattern of the leaves of the downy rattlesnake orchid, supposedly resembling snakeskin, probably led to its use as a cure for snakebite. Indians used the plant to treat snakebite, toothache, colds, kidney problems, loss of appetite, eye ailments, burns, sores and swollen lymph nodes. So far, none of these uses have proved valid.

large twayblade

Liparis liliifolia

Blooms in May and June

This orchid ranges from Minnesota and Ontario east to New Hampshire, and south to Oklahoma and Georgia. It is of conservation concern in a few northern states.

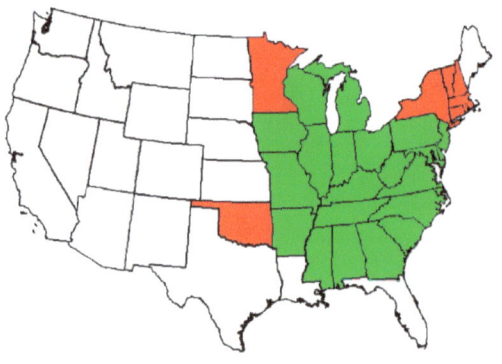

In Arkansas it is found in scattered areas in the northern half of the state.

The large twayblade grows in moderate shade in forests. In Lovell Hollow it has been observed on the old logging road in the middle elevations, and at Heartsong it has been found in the forest at the edge of the clearing around the green Man statue.

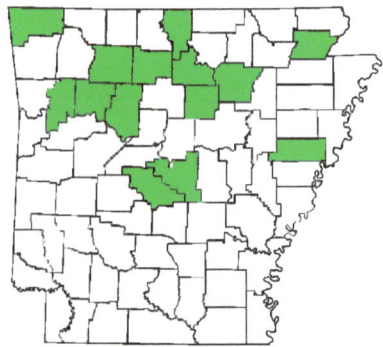

This lovely orchid is easily overlooked even when in bloom, as the mauve colored flowers blend into the shadows.

The name 'twayblade' refers to the two broad, shiny leaves. Other names sometimes used include lily-leaved twayblade, and the unlikely 'mauve sleekwort.'

green adder's-mouth orchid

Malaxis unifolia

Blooms in May and June

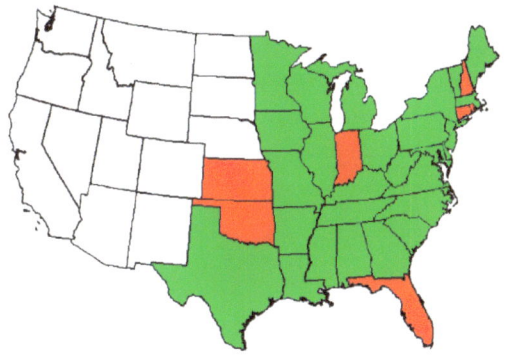

The green adder's-mouth orchid is found over the eastern half of the United States, but is on the list of species of conservation concern for several states.

In Arkansas it is found scattered throughout the state in shaded moist woods and bogs.

Here it is found in the woods along the trail to the river, around a boggy seepage, which crosses the trail. This clump is the first record of the species reported for Newton County.

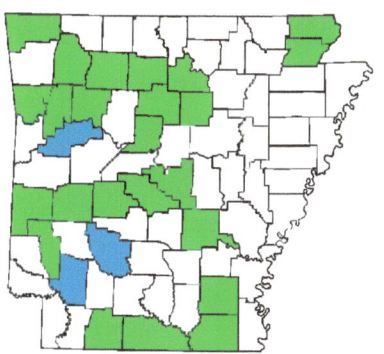

The plant stands about 6 inches tall and has a stem of 20 to 50 tiny green flowers. A single leaf wraps around the stem.

nodding ladies'-tresses

Spiranthes cernua

Blooms in November

This orchid is found throughout the eastern and midwestern US, and is on the species of conservation concern list in at least one state. In Arkansas it has been recorded in nearly every county. Here it may be encountered in cleared or mowed areas on the ridge.

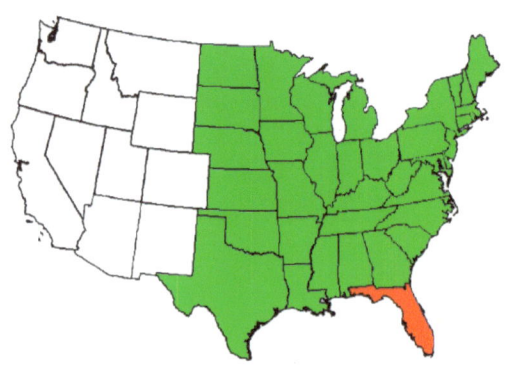
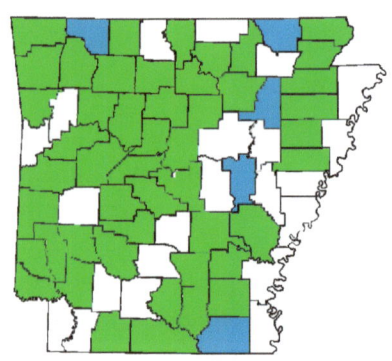

This flower is one of several species of very similar appearing orchids which have tiny, usually white flowers spirally arranged around the stem, as though curled like a lady's hair. The species vary slightly in color, fragrance, blooming season, and whether the leaves are still present when blooming, in addition to more technical characters.

The nodding ladies'-tresses have white flowers with white throats, leaves absent at blooming, and may be lightly fragrant.

This species is very variable and interbreeds with several other species where their ranges cross, and incorporates their characteristics. Fortunately the other compatible species are not found here, so it is more easily identified.

nodding ladies'-tresses, *Spiranthes cernua*

slender ladies'-tresses

Spiranthes lacera var. *gracilis*

Blooms in August and September

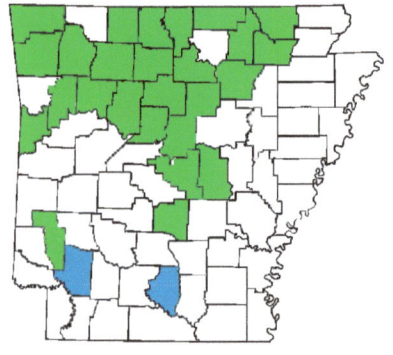

This orchid is found from Nebraska to Maine and south to Texas and Florida. It is on the species of conservation concern list of at least one state.

In Arkansas it is found mostly in the Ozark and central regions.

Here it is especially abundant in mowed and cleared areas and sunny paths throughout the camp.

Small white flowers spirally wound around the stem characterize this orchid, like other species of ladies'-tresses. It can be distinguished from the other ladies'-tresses found here by its blooming season, mostly September, and especially by the green color in the throat of the flower.

little ladies'-tresses

Spiranthes tuberosa

Blooms in August and September

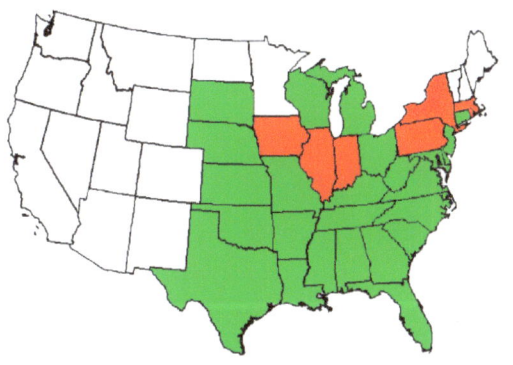

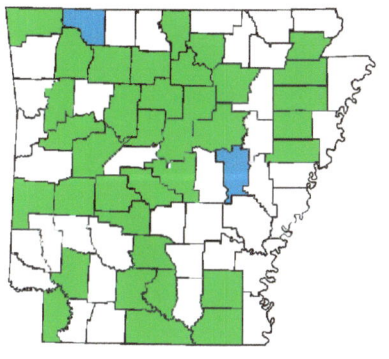

This ladies'-tresses orchid is found from Michigan to Massachusetts and south to Texas and Florida. It is on the species of conservation concern list of several states.

In Arkansas it is found in scattered locations throughout the state.

Here it may be found in lawns and cleared or mowed areas or along paths throughout the ridge.

One of the smallest orchids in Arkansas, it is slightly smaller than the other ladies'-tresses orchids. It stands from a foot to 18 inches tall and has a spiral of all white flowers. It blooms in August and September. It is found in the same areas as the slender ladies'-tresses, but is just ending its blooming season as the other begins.

spring ladies'-tresses

Spiranthes vernalis

Blooms in July

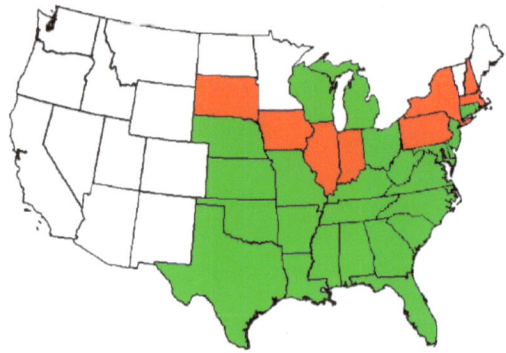

This orchid is found from South Dakota to New Hampshire and south to Texas and Florida. Although it is on the list of species of conservation concern in several northern states, it is one of the most common and widespread orchids in the southern states.

In Arkansas it is found in nearly every county, although it had not been recorded in Newton County before these specimens illustrated here. A few have been found in mowed or cleared areas on the ridge.

The name *vernalis* implies spring blooming, but here the blooming is in July. In the southern states it blooms much earlier, and it is still the earliest of the ladies'-tresses in this area.

crane-fly orchid

Tipularia discolor

Blooms in August and September

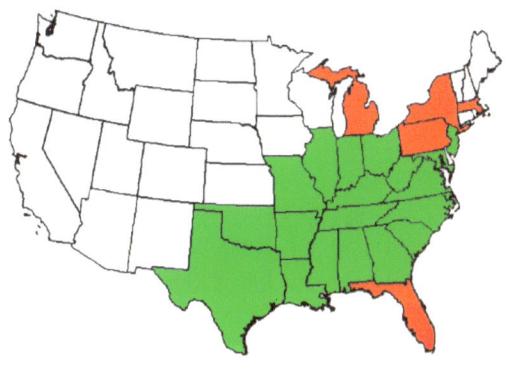

This orchid ranges from Michigan to Massachusetts and south to Texas and Florida.

In Arkansas it is found scattered mostly in the central and southern counties.

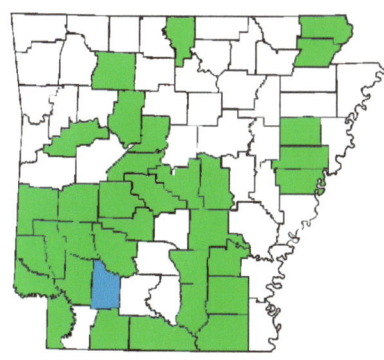

Here it has been found in the woods next to the river. This site was the first recorded for Newton County.

This is one of several species of orchids, which are most visible in winter. Its leaves appear late in the year and remain green throughout the winter, while the leaves are gone from the trees. Its attractive foliage is a vivid purple on the underside, and it has purple spots on the top, making it easily recognized.

The lopsided flowers are tiny, reddish, and fancied to resemble crane flies clustered on the stalk.

It takes a bit of imagination.

19

three birds orchid

Triphora trianthophora

Blooms in August and September

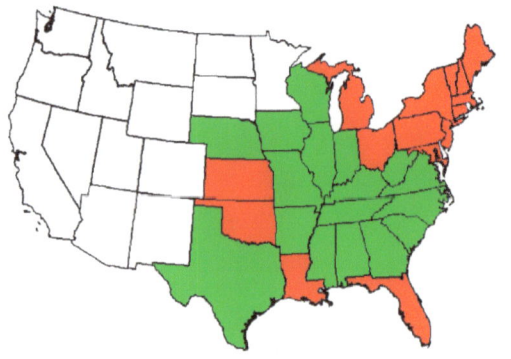

The three birds orchid is found from Nebraska to Maine and south to Texas and Florida. It is on the species of conservation concern lists of most of the northeastern states.

In Arkansas it is found in the Ozarks and Ouachita mountains.

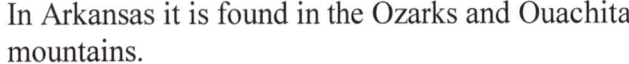

Here it may be found in the woods from the river to midway up the ridge.

The name 'three birds' alludes to the trait of often blooming in threes, with the flowers pointing upward like hungry baby birds.

This is one of the most interesting orchids of the area, in addition to being strikingly pretty. It is easily overlooked because the flowers are only open fully for one day. All the plants in a given area (sometimes said to cover several states) will blossom at the same time.

There will be several flushes of blooms within a season, but the triggering mechanism has not been adequately explained.

Most theories deal with cool nights or temperature differences between day and night, but even sheltered plants may bloom with others in the area. The writer has observed that cut buds kept in water in the house bloom with those in the woods a half-mile away.

three birds orchids spend most of their lives underground and send up the succulent foliage only during blooming season. They may remain underground for years without blooming.

Their coordinated blooming is apparently effective in achieving pollination, as most flowers will be observed to have set seed.

Plants also reproduce by offshoots from the roots, and most plants will be seen in clumps of clones.

When and Where to Look

First it should be remembered that even though none of the orchids outlined here are on an Arkansas list of conservation concern, all of them are endangered somewhere in the United States. All of them are protected in the Upper Buffalo Wilderness Area.

Some of the species listed are locally common, particularly *Spiranthes lacera* var. *gracilis*, the slender ladies'-tresses, and *Spiranthes tuberosa*, the little ladies'-tresses. These appear in abundance in the lawns and paths on the ridge in late summer, and can be easily found before mowing occurs. *Goodyera pubescens*, the downy rattlesnake orchid, is common through the woods all the way to the river, and is common in the bottomland forests as well. The coralroots may be common in the bottomlands, but are usually overlooked because they are so inconspicuous, whereas the lady's-slippers are not common, but are much more conspicuous. three birds orchids, *Triphora trianthophora*, is common from the middle woods to the bottomlands, but is not likely to be seen except on the 4 or 5 days a year when they all bloom at once. The Showy orchis, *Galearis spectabilis*, is fairly abundant in the bottomlands on the other side of the river, but bloom when the river is high.

There is no time of the year when orchids can't be recognized somewhere in the area, by their foliage if not flowers. Even a walk in the winter may reveal orchids if the ground is not covered in snow, as the downy rattlesnake orchid, *Goodyera pubescens*, the crane-fly Orchid, *Tipularia discolor*, and Adam-and-Eve, *Aplectrum hyemale*, emerge in October and are green all winter long, and are actually most easily located then.

The following chart indicates what species you might encounter throughout the year if you are observant.

	Flowers	Foliage (in winter)
January		*Tipularia discolor* *Goodyera pubescens* *Aplectrum hyemale*
February		*Tipularia discolor* *Goodyera pubescens* *Aplectrum hyemale*
March		*Tipularia discolor* *Goodyera pubescens* *Aplectrum hyemale*
April	*Cypripedium pubescens*	*Tipularia discolor* *Goodyera pubescens* *Aplectrum hyemale*
May	*Aplectrum hyemale* *Corallorhiza wisteriana* *Cypripedium pubescens* *Galearis spectabilis* *Liparis liliifolia* *Malaxis unifolia*	
June	*Liparis liliifolia* *Malaxis unifolia*	
July	*Goodyera pubescens* *Spiranthes vernalis*	
August	*Tipularia discolor* *Corallorhiza odontorhiza* *Goodyera pubescens* *Spiranthes lacera* *Spiranthes tuberosa* *Triphora trianthophora*	
September	*Corallorhiza odontorhiza* *Spiranthes lacera* *Triphora trianthophora*	
October	*Spiranthes cernua*	*Tipularia discolor* *Goodyera pubescens* *Aplectrum hyemale*
November	*Spiranthes cernua*	*Tipularia discolor* *Goodyera pubescens* *Aplectrum hyemale*
December		*Tipularia discolor* *Goodyera pubescens* *Aplectrum hyemale*

Index

Adam-and-Eve 1, 22
Adder's-mouth
    green 13
Aplectrum
    hyemale 1, 22, 23
Autumn coralroot 3, 4
Buffalo River iii
Corallorhiza
    odontorhiza 23
        var. *odontorhiza* 3
        var. *pringlei* 4
    wisteriana 5, 6, 23
Coralroot .... iv, 3, 22
    autumn 3, 5
        Pringle's 4
    spring 5, 6
    Wister's 5, 6
Crane flies 19
Crane-fly orchid 19, 22
Cypripedium
    parviflorum
        var. *parviflorum* 8
        var. *pubescens* 7, 8, 23
Downy rattlesnake orchid .. iv, 10, 11, 22
Galearis
    spectabilis 9, 22, 23
Goodyera
    pubescens 10, 11, 22, 23
Green adder's-mouth 13
Ladies'-tresses iv
    little 17, 22
    nodding 14, 15
    slender 16, 17, 22
Lady's-slipper .... iv, 22
    large yellow 7, 8

Large twayblade 12
Large yellow lady's-slipper 7, 8
Lily-leaved twayblade 12
Liparis
    liliifolia 12, 23
Little ladies' tresses 17
Lovell Hollow .... iii, iv
Malaxis
    unifolia 13, 23
Mauve sleekwort 12
Nodding ladies'-tresses 14, 15
Pringle's autumn coralroot 4
Putty-root 1
Rattlesnake orchid
    downy 10
Showy orchis 9, 22
Slender ladies'-tresses 16, 17, 22
Spiranthes
    cernua 14, *23*
    lacera
        var. *gracilis* 15, 16, 22, 23
    tuberosa 17, 22, 23
    vernalis .... iv, 18, 23
Spring coralroot 5
Three birds orchid .... iv, 20, 21, 22
Tipularia
    discolor 19, 22, 23
Triphora
    trianthophora 20, 22, 23
Twayblade
    Large 12
    Lily-leaved 12
Upper Buffalo Wilderness Area iii
Wister's coralroot 5, 6

www.ingramcontent.com/pod-product-compliance
Lightning Source LLC
Chambersburg PA
CBHW041306180526
45172CB00003B/985